D1532889

SCIENCE TO THE RESCUE
ADAPTING TO CLIMATE CHANGE™

ADAPTING TO
PLANT AND ANIMAL
EXTINCTIONS

KATHY FURGANG

NEW YORK

For Steven

Published in 2013 by The Rosen Publishing Group, Inc.
29 East 21st Street, New York, NY 10010

Copyright © 2013 by The Rosen Publishing Group, Inc.

First Edition

All rights reserved. No part of this book may be reproduced in any form without permission in writing from the publisher, except by a reviewer.

Library of Congress Cataloging-in-Publication Data

Furgang, Kathy.
Adapting to plant and animal extinctions/Kathy Furgang.—1st ed.
    p. cm.—(Science to the rescue: adapting to climate change)
Includes bibliographical references and index.
ISBN 978-1-4488-6850-6 (library binding)
1. Extinction (Biology) 2. Climatic changes—Environmental aspects. I. Title.
QH78.F87 2013
591.3'8—dc23

                                                    2011050340

Manufactured in the United States of America

CPSIA Compliance Information: Batch #S12YA: For further information, contact Rosen Publishing, New York, New York, at 1-800-237-9932.

On the cover: Polar bears face a highly uncertain future as the Arctic ice they depend upon for travel, hunting, and food is melting away at a rapidly increasing rate.

CONtents

INTROduction

Recently, polar bears have come to symbolize a problem with Earth's climate. You may have seen the photos before. These strong yet vulnerable creatures are stranded on tiny blocks of ice instead of spread out on sprawling sea ice in the Arctic. Scientists have warned that the rapidly melting sea ice in the Arctic could bring about the extinction of this beautiful mammal.

Climate researchers have explained that the average surface temperatures on Earth have been steadily climbing over the past century. Although the temperature increases are small and seem insignificant to us, scientists say that they are having a profound effect on the planet. For the polar bear and many other animals and plants, it means that populations are becoming endangered and extinct. Scientists say the changing climate is causing the rate of extinction to be one hundred to one thousand times greater than normal. That rate works out to be an extinction of a plant or animal species every twenty minutes. At the current rate of extinction, more than one million species will be lost forever by the year 2050. Even if action is taken now, many of those organisms

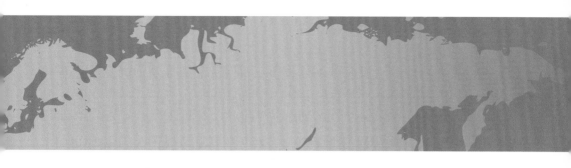

are unlikely to be able to be saved because their population numbers are already too low.

In the case of the polar bear, the problem is that the sea ice in the Arctic is melting faster than expected. Ordinarily, polar bears roam for miles on the ice. They depend upon it to find their food because the ice connects areas of land that they use for hunting. With less ice to roam on, polar bears must swim greater distances between hunting

This polar bear moves across sheets of ice called ice floes. As ice floes in the Arctic shrink and melt, the polar bear has less room to roam and hunt for food.

grounds in rough, icy waters. They risk drowning. In addition, their food supplies are also decreasing. The animals they depend on for food have begun to relocate because the changing climate has affected them as well. Polar bears are not adapted to warmer temperatures, and they are beginning to starve because they cannot get the food they need to live.

Animals and plants depend upon each other for food and nourishment. When just one part of this food web is altered, is damaged, or disappears, the entire food web is affected. The healthiest ecosystems have the most biodiversity, which is the variety of life in an area. Extinction is the greatest danger to biodiversity and to food webs. Scientists say that there were six great periods of mass extinction on Earth and that we may be entering a seventh one now due to global warming and climate change.

While the data on polar bears is devastating, humans can still do a lot to save threatened and endangered species all around the world. The polar bear can be helped if humans take action now. Learning about the causes and effects of climate change can help us to save the planet and the plant and animal species that live on it. Species all around the world will almost certainly be suffering in the near future, but those that are particularly adapted to colder climates are already suffering. That is because there is nowhere cooler on the planet for them to relocate to. As we learn more about climate change and its effects on our planet, we can figure out how to best help plant and animal species—including the human species—to survive.

CHAPTER one

What Is Causing Climate Change?

When scientists say that countless plants and animals are in danger of becoming extinct in the coming years because of global climate change, it is only natural for humans to want to help. They also want to know what caused this problem in the first place. According to scientists, it is human activity that has caused an overall change in Earth's climate.

THE INDUSTRIAL REVOLUTION

Human-driven climate change greatly intensified with the onset of the Industrial Revolution. This was a time when nations such as the United States and Great Britain began to invent ways to get work done with machinery instead of by hand. The end of the eighteenth century and the beginning of the nineteenth century saw a huge increase in the number of factories. The factories made goods faster, cheaper, and easier than ever. Steam-powered factories could work around the clock. Steam-powered trains transported goods to locations throughout newly industrialized nations. The burning of coal created the steam that powered all this heavy machinery.

A revolution had begun. More people began working in factories than on farms. Soon, gasoline-fueled cars were manufactured on assembly lines and became easier and cheaper to produce. Over the years, more families were able to buy cars. They could now travel faster and more comfortably than with a horse and carriage, so car sales climbed higher and higher. People saw the Industrial Revolution as progress. In many ways it certainly was progress and made our lives much easier. However, there were also drawbacks to this new Industrial Revolution. These drawbacks took a bit longer for people to recognize than the obvious benefits. The factories, machines, electricity, and transportation all required the burning of fossil fuels.

A fossil fuel is a resource extracted from the earth that has taken a very long time to form. Coal and gas are examples

Beginning in the nineteenth century, the Industrial Revolution caused an abundance of air pollution in Earth's atmosphere for the first time in human history.

of fossil fuels. Coal has formed over millions of years from the remains of once-living organisms. Burning coal provides the energy needed to run a factory, heat a home or business, and provide electricity to cities or towns. At the beginning of the twentieth century, fossil fuels were being burned and consumed in large quantities for the first time. When fossil fuels are burned, they release pollution and noxious gases into the air. This has always been apparent when looking at old factories and mills from the early twentieth century. The black smoke and pollution released into the air only increased over the years.

The introduction of the automobile on a large scale only made things worse. Gasoline is another fossil fuel that was burned and released pollution into the atmosphere. Just one hundred years ago, there were not even enough paved roads to connect every town in the United States. Today, many families own more than one car, and highways become jammed for miles due to the high volume of traffic.

Scientists say that it is the burning of fossil fuels that has caused the overall increase in temperatures over the century. Why? Because within the pollution released by the burning of fossil fuels are so-called greenhouse gases. These gases include carbon dioxide, water vapor, methane, and ozone. It is the increase in carbon dioxide that scientists are singling out as the single greatest contributing factor to the climate change problem that our planet is experiencing.

FACTS ABOUT CARBON DIOXIDE

Carbon dioxide is a natural gas that occurs in the atmosphere. In fact, you release carbon dioxide out of your body every time you exhale your breath. It is a waste gas that is produced naturally by animals after they process the oxygen their bodies need.

Trees and other plants use carbon dioxide to drive photosynthesis. Plants couldn't survive without the carbon dioxide in the air. So what makes the carbon dioxide produced by burning fossil fuels so dangerous? Scientists say that it is the unnaturally large amount that has accumulated in the atmosphere over the past century. Large-scale burning of fossil fuels is a human activity that had not existed before the past two centuries, and that is the same period of time in which global temperatures have been increasing steadily and fairly dramatically. Scientists have found that carbon dioxide and other greenhouse gases are accumulating in the atmosphere and creating a virtual blanket over the atmosphere. The blanket traps heat, much like the sun's heat becomes trapped under the panes of glass in a greenhouse.

The greenhouse effect is natural and even necessary. It is what makes the planet temperate enough to be inhabitable. If the sun's heat were not trapped in the atmosphere, the

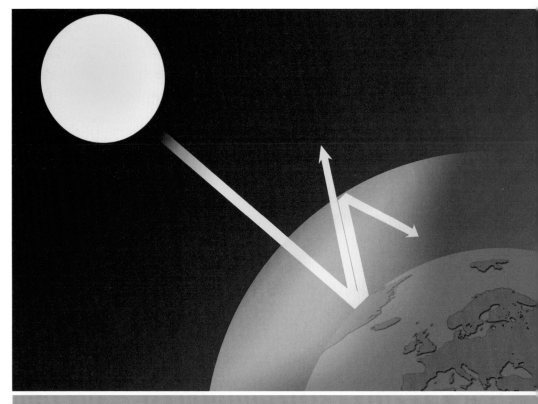

This diagram of the greenhouse effect shows how some of the sun's radiation is absorbed by Earth, some is reflected back into space, and some becomes trapped by the gases in the atmosphere and re-radiates back to Earth.

planet would be far too frigid to support life. It is the excessive buildup of these greenhouse gases, however, caused primarily by the burning of fossil fuels, that is causing global warming and resulting in climate change. Scientists have recorded an increase in global temperatures of 1 degree Fahrenheit (0.55 degrees Celsius) over the past one hundred years. This may sound like a very small amount. However, the amount is quite significant over such a relatively short period. In addition, it is an average, overall temperature increase worldwide, which means that some areas have seen a greater increase than one degree. Scientists have been able to attribute the increase in global temperatures to the increase in carbon dioxide emissions. They have noticed a direct correlation with the increase of carbon dioxide in the atmosphere and the increase in Earth's global surface and ocean temperatures.

UPSETTING THE BALANCE

How does global climate change affect living things? There are plenty of ways. The most significant is that it ruins plant and animal habitats. The threat is so great that scientists claim that 25 percent of all plant and animal species on Earth are threatened with extinction by 2050.

Our planet is filled with distinct and varied ecosystems around the globe. Most plants and animals are adapted specifically to live in their particular habitat and ecosystem. Polar bears live in the Arctic region because their bodies require the temperature and climate conditions of the area. Lizards and cacti plants are adapted to hot deserts because of the temperatures, climate conditions, and amount of water in the region. The balance between living things and their environment can be fragile. Too much water in a desert can cause many organisms to die. Higher temperatures in the Arctic can also cause organisms to die. Over time, they may not be able to adapt to the temperature changes or their food supplies will be destroyed.

For example, seals are one of the polar bears' main food sources. In turn, seals depend on fish for their own food supply. If the temperatures of the Arctic change so much as to kill fish off or make them move to another area in search of cooler water and more plentiful food supplies, then the food chain is disturbed. The seals cannot find the food they need to survive. In turn, the polar bears cannot find the food they need. If animals cannot adapt to meet their needs in this changing environment, they must either move to a new environment or die. Because of diminishing food supply and diminishing pack ice, polar bears are being forced to swim longer distances to find the food they need to live. Many are now drowning in the process.

The polar bear is just one example of the stresses placed upon plant and animal species by climate change. Many other animals are being forced to adapt, move, or die. According to the National Science Foundation, the entire ocean environment is being altered by the effects of climate change. Tiny ocean organisms called plankton produce oxygen and remove carbon dioxide from the water by using it in photosynthesis. When warmer temperatures cause these organisms to die off, the ocean becomes much more acidic because carbon dioxide is no longer being removed by plankton. Plankton is also an important food source for fish and whales. In addition, increasingly acidic oceans are killing off the coral reefs that provide sustenance and shelter to many ocean species. According to a study reported by *Science News*, coral have been migrating north over the past eighty years. This is affecting the food chain and biodiversity of the regions they once inhabited.

THE HUMAN ROLE IN EXTINCTIONS

The effect of human activity on plant and animal extinction is not a new story. Human activity has affected plant and animal life throughout history. It is estimated that two thousand species of birds from the Pacific Islands have become extinct since humans began to colonize the land and use its resources. That's about 15 percent of the total population of birds on Earth. Overfishing is another human activity that has caused extinctions. Many of the mussel, clam, and fish species have disappeared in the last century around North American waters.

When humans build towns and cities, plant and animal habitats are destroyed. Trees are leveled, and paved roads run through the ecosystems that animals rely on for food. If a species of plant in an area is already suffering or endangered, construction and development projects may wipe the organism away entirely. That will of course affect the way animals within that ecosystem obtain their food. Animals such as raccoons, foxes, or rabbits must relocate to find alternative or additional food sources, or they must adapt to the new environment.

Human activity such as building housing developments and roads puts plant and animal habitats at risk, shrinks their territory, and disrupts the food chain.

ENDANGERED VS. EXTINCT ORGANISMS

We use the terms "extinct," "endangered," and even "threatened" when we talk about the status of different species. What do these terms mean? The most severe of these terms, of course, is "extinct." This means that there is no more of that particular species left on Earth; the species is gone forever. Every last organism of its kind has died off worldwide.

A GLOBAL RESPONSE TO GLOBAL WARMING

Since 1992, environmental and political leaders have been trying to decrease the amount of greenhouse gases in the atmosphere. Political help is needed to make policy changes that will limit the amount of greenhouse gas emissions that factories, cars, and nations as a whole can produce. Leaders of various nations have voluntarily reduced their carbon emissions. In January 2010, the United States pledged to reduce its greenhouse gas emissions by 17 percent by the year 2020. Individual states are also doing their part. The Regional Greenhouse Gas Initiative is an agreement among ten Northeastern states to voluntarily cap greenhouse gas emissions, improve the environment, and help curb the effects of climate change.

An endangered species is threatened with extinction throughout all of its range or a large portion of its range. There are currently more than fifteen thousand plant and animal species that are considered endangered. According to the World Wildlife Fund, the ten most endangered animals on Earth include the tiger, polar bear, Pacific walrus, Magellanic penguin, leatherback turtle, bluefin tuna, mountain gorilla, monarch butterfly, Javan rhinoceros, and giant panda. Some of these species are close to the brink of extinction. The Javan rhinoceros, for example, has only two main population groups worldwide, totaling fewer than sixty individuals. Endangered plants include many species of begonia, palms, orchids, the pitcher plant, and thousands

The world's few remaining mountain gorillas live in national parks. This mother and baby gorilla are in Volcanoes National Park in Rwanda.

of other plants, many of them existing in rain forests that are themselves endangered. Rain forests around the world have been destroyed by human activity for many years. These areas happen to be the home of many rare plants that can be used in medicines and to save human lives.

Another term exists for animals that are close to becoming extinct. A threatened species is one that is in danger of being added to the endangered list. An animal may be listed as threatened in a particular range or throughout the world. Organisms are tracked by the World Conservation Union. This group categorizes each plant and animal species using a rating system that indicates how threatened or endangered they are.

While human-driven habitat destruction is the main reason why so many animals and plants are on the endangered and threatened species lists, humans can also do a lot to help these animals return from near extinction. The wolf population, for example, has been able to increase sufficiently to be removed from the list. One thing that has helped has been the Endangered Species Act of 1973. This law made it a crime to kill any plant or animal on the endangered species list. Humans have also been prevented from developing areas where endangered plants and animals live.

With the negative effects that climate change brings, the struggle of threatened and endangered species gets even harder. Their conservation will become even more important in the coming years as the effects of climate change become more severe.

CHAPTER two

What the Experts Predict

Scientists have been keeping temperature and climate data for over a century, and they have also been looking ahead to predict what effect our current activities will likely have on the changing climate in the future. Mathematical climate models were invented in the 1950s and have improved greatly over the decades. Today's models take into account the effects that air circulation over oceans has on the climate. When this data can be input into a computer model, scientists can figure out the mean overall temperatures for certain areas around the globe. These models can also take

How the World Will Heat Up

temperature changes compared to the average value in the years 1980-1999 (medium scenario for climate change)

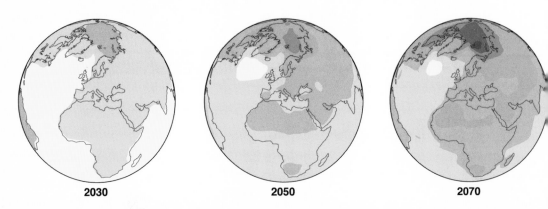

2030	2050	2070

Computer models predict temperature changes in future decades. The darkest areas of each map indicate the areas where the greatest temperature increases can be expected.

into account different scenarios for future levels of carbon dioxide and greenhouse gas emissions.

COLLECTING DATA, DRAWING CONCLUSIONS

One of the most important international organizations studying the problem of global warming and climate change is the United Nations World Meteorologists Organization. It gets the help of over eighty-five countries, each contributing to a cooperative study drawn from over ten thousand observation sites on land, seven thousand at sea, and ten from satellites in space. The data that is collected helps to refine computer climate models

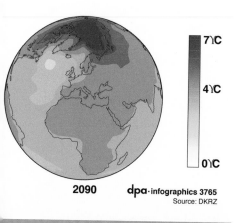

7°C

4°C

0°C

2090   dpa·infographics 3765
Source: DKRZ

that can better predict how global temperatures will change over the next several decades.

These computer models can compare differing amounts of greenhouse gases to see exactly how more or less emissions would impact temperature and climate in the future. For example, what if the amount of greenhouse gas emissions was cut dramatically over the next decade? What effect would that have on the climate ten, twenty, or even fifty years from now? According to most models, scientists say that even with a significant decrease in greenhouse gas emissions into the atmosphere, there will still be an overall increase of 3 to 10°F (1.7 to 5.6°C) by the year 2100. The increase would be greater than what we have already seen so far because the greenhouse gases that exist in the atmosphere already dissipate very slowly and linger for a long time.

THE FUTURE PROSPECTS FOR PLANT AND ANIMAL SPECIES

What does this data mean for plant and animal species? A study was published in 2004 by the Intergovernmental Panel on Climate Change (IPCC) that used the results of average global warming prediction models. The study took two years to put together and was the first major assessment of the effects of climate change. It found that 15 to 37 percent of all species in the regions studied could be extinct or extremely endangered by 2050. Six distinct regions with different types of climates and biodiversity were studied, representing about 20 percent of Earth's land surface. But what if these results were applied worldwide? That would mean that over a million species are currently threatened by climate change.

The study included interesting predictions about changes to animal habitats and the likelihood of survival for various species. Species in mountainous areas such as those in Europe, Central and South America, Australia, and South Africa would survive best. This is because the animals could move uphill and seek cooler temperatures when their current habitat became unable to support their needs. Desert animals are likely to have trouble surviving because their habitat is so flat. Migrating to a cooler climate would require moving over vast distances. Out of the 1,870 species that were examined in the study, one-third of them were not expected to survive the conditions that current climate change models predict over the next forty or fifty years.

Birds are thought to have a good chance of survival because they can reach new and hospitable habitats by flying. However, they do require certain plants and trees for shelter and nourishment, so they would have a major challenge if food chains and food webs began to break down.

Butterflies in Australia were part of the IPCC study. Out of the twenty-four species studied, only three were predicted to be able to survive the likely climate changes in their current climate zones. The study predicted that about half of the butterfly species would become extinct because of an inability to adapt to the new conditions.

Some ecosystems support unique plant and animal life. In the Brazilian savannah, there are many plants and trees that do not exist anywhere else in the world. Scientists predict that up to 48 percent of these species would become extinct. The savannah covers one-fifth of the country of Brazil.

The predictions in Europe were not as devastating, but, even there, up to 17 percent of the plant species were predicted to become extinct by 2050. About 25 percent of the bird species could become extinct, including some species that are already rare. The Scottish crossbill, for example, is found only in Scotland and is expected to be unable to survive the likely changes to its environment. It would also be threatened by other species relocating there from warming climates and providing additional competition for resources.

In order to adapt to these changes, scientists predict that animals will have to change some of their behaviors. The warmer winters will cause some animals to stay active all winter instead of hibernating to avoid the winter chill and scarcity of food. That means there will be more animals competing for scarce food throughout the entire winter. This change will bring additional challenges to a food web that will already be straining under increasingly scarce resources.

Clearly, climate change is not the only pressure that threatens the continued existence of vulnerable plant and

Some animals adapt better than others to temperature changes. These wild horses from the Namib Desert have survived for over one hundred years where temperatures often reach 113°F (45°C).

animal species, but it is the one that drives most of the others. Animal species will relocate to new places to try to adapt to climate changes. But when the areas that species are relocating to are already degraded or destroyed by human activity such as development and pollution, the animals are forced to compete for less and less space and fewer food and water resources. Add to this the widespread habitat loss that is already occurring due to climate change, and the stress may be too much for many species. Trees are dying in forests due to drought, disease, and warming temperatures. Many animal species that rely on these trees have trouble surviving in these ecosystems that are already suffering.

INVASIVE SPECIES

According to *An Inconvenient Truth*, written by former U.S. vice president Al Gore, the warming of Earth is not evenly distributed. The Arctic and Antarctic regions are experiencing an even more pronounced temperature change than the tropic, subtropic, temperate, and boreal regions. When scientists predict an overall global temperature increase of 5°F (2.8°C), climate models clarify that this means about 1°F (0.6°C) at the equator, but about 12°F (6.7°C) at the North and South Poles. These changes can have devastating effects on the organisms in these regions.

According to climate predictions even less extreme than this one, animals will be forced to shift the ranges that they once roamed and non-native plants better acclimated to new climate conditions may begin flourishing. That means they will be invading new areas where they had never lived before. In some ways, this means an increase in biodiversity in a region. However, the invasive species can have very negative and unpredictable effects upon food chains. According to the National Wildlife Federation, about 42 percent of the threatened or endangered plant and animal species on Earth are at risk because of invasive species in their environment.

A new species in a region can end up reproducing too quickly and successfully because it does not have natural predators in its new home. That may cause too much competition for food, water, and territory for the native species, who may begin dying off. The food web of the ecosystem

This bighead carp is an invasive species from Asia that has wreaked havoc on native species in Wisconsin's St. Croix River.

becomes out of balance. For example, if lake trout that normally live in the Great Lakes are somehow relocated to another lake, such as Wyoming's Yellowstone Lake, they are an invasive species in that area. They begin competing with the native fish in that lake, including the cutthroat trout. The presence of the lake trout could seriously endanger the entire cutthroat trout population.

Invasive animal species are not the only problem that a changing ecosystem faces. Invasive plants can also be a problem. Seeds may end up in a new environment in several ways. Often migrating animals bring seeds with them that are then relocated and germinate in new territory. These seeds often travel after being stuck to the fur of some animals or being eaten and excreted as waste by the animals.

TAKING A CLOSE LOOK AT THE OCEAN

When scientists look at ocean food chains and how climate change may affect them, they keep coming back to the organisms at the bottom of the food chain—plankton. These tiny organisms provide the food that starts the energy moving through the food chain. Studying them may produce the next big solution to the climate change problem. Scientists are studying the effects of temperature changes on plankton. They will also study plankton to see how it reacts to more acidic water. When more carbon dioxide is dissolved in seawater, the pH of the water is reduced and carbonic acid is produced. The study of how increasingly acidic seawater affects plankton—the base of all ocean food chains—is of vital importance to every organism within the global food chain, humans very much included.

If the invasive seeds germinate and the resulting plants grow quickly and crowd out native plant species, the same food web imbalance may occur.

Not only do invasive plant and animal species cause extra competition for resources in an area, they may also spread disease or attack the native species in some way, preventing them from reproducing. Even bacteria and fungi can become invasive in an area, just like other plant and animal species. Many bacteria and fungi thrive where the temperature is warmer. For this reason, more and more of the planet is becoming hospitable territory for them as global temperatures rise. The consequences can be catastrophic to vulnerable species. In 2006, a scientist at Monteverde Cloud

This head of coral shows the effects of coral bleaching. The rising ocean temperatures cause algae to die, which in turn creates this whitening effect on the coral and threatens its survival.

Forest Reserve in Costa Rica reported that about two-thirds of the harlequin frog species in the area had been killed by a fungus that caused disease in the frog.

WARMING AND WATER HABITATS

For some species, temperature fluctuations, or swings, are more dangerous than consistently rising temperatures. This is especially true for ocean life. Coral, for example, cannot survive extreme temperature changes and would adapt better to a slow but constant rise in the overall temperatures of ocean water.

One problem in the ocean that scientists attribute to climate change is coral bleaching. Coral polyps live in the ocean with colorful algae. The algae provide the polyps with the oxygen and nutrients they need to live. Because of temperature fluctuation, however, the algae has been dying off without being replaced. This, in turn, causes the coral to die off. As with all other food webs, if the coral continues to die off, the fish and other organisms that rely on the coral reef for survival will also be threatened.

CHAPTER three

The Consequences of Plant and Animal Extinctions

Many people may hear about the increased number of plant and animal extinctions caused by climate change and feel sad for the species that are affected. But unfortunately, the problem is even larger than it first appears, and our own survival as a species is at stake. Humans rely on natural ecosystems just as other animals do. Although we tend to make some of our own ecosystems look somewhat unnatural with glass and steel skyscrapers, concrete roadways, and other human-made objects, we owe a lot to the natural world and depend upon it for our survival.

THE HUMAN TOLL OF SPECIES LOSS AND KNOWLEDGE GAPS

The natural world is responsible for helping us make about half of our medicines. For example, aspirin used to be made from the willow tree. Antibiotics are made from fungi and are used routinely to treat countless illnesses. A periwinkle flower from Madagascar is used in medicines to treat leukemia, a serious blood disease. The chemotherapy drug Taxol was developed from the Pacific yew tree and is used to treat lung and breast cancers.

A researcher collects a sample of Pacific yew bark, which is used to make certain cancer medicines.

Even the bite of a snake can show how humans can benefit from an ecosystem kept in balance. The South American pit viper, called *Bothrops jacara*, has a substance in its venom that can be used to lower human blood pressure. The substance has been used in medicines for years and has saved many lives. A particular fungus from the soil on Easter Island is used in medicines to treat people who have received organ transplants or undergone heart surgery. The fungus is even known to extend the lives of mice. Studying plants and animals can help us learn a lot about disease.

Many of these medicinal substances are now re-created in laboratories on a large scale, but without finding these things in nature, we would never have discovered their positive effects on human health. The natural world still contains untold numbers of as yet undiscovered future cures, medicines, and treatments, assuming the world's ecosystems can survive intact in the coming decades.

With the increased number of organism extinctions, there will be a loss of scientific knowledge from our natural world. According to the National Geographic Society, scientists have still not discovered or classified over 86 percent of the species on Earth. They predict that the planet is home to 8.7 million species, but they have been able to catalog less than 15 percent of these species, which is just about 1.2 million. They predict that many of the unrecorded species will become extinct before they can ever become cataloged. With extinction rates now ten to one hundred times their natural rate, many of those species will never be discovered, nor will their potential benefits to human health.

NOTABLE EXTINCTIONS

Here are just a few extinctions that have taken place over the past couple of decades. These species will no longer be seen on Earth because every last one of their kind has died. Most of these species are extinct because of the results of human activity.

- 1982: The Tecopa pupfish lived in the hot springs of the Mojave Desert. Habitat loss and construction by land developers caused this animal to become extinct.
- 1989: The golden toad lived in high altitudes in Costa Rica but became extinct because of pollution, climate change, and disease.
- 1996: The Zanzibar leopard was last seen on an island off of Tanzania. There were thought to be a couple of sightings over the years, though none could be confirmed.
- 2000: The last population of Pyrenean ibex lived in Spain and became extinct due to hunting.
- 2004: The po'ouli was also known as the black-faced honeycreeper. It was native to Maui, Hawaii, and was not discovered until the 1970s. There were only three left by 1997, and preservation efforts failed to save the bird. Disease, predators, and habitat and food source loss caused the animal's extinction.
- 2006: The West African black rhinoceros became extinct due to poachers, who illegally hunted the animal on protected land. They killed the rhinos for their horns.

- 2007: The Madeiran large white was a butterfly that lived in forests of the Madeira Islands in Portugal. Habitat loss, pollution, and fertilizers killed the last of this butterfly population.

Plants that have become extinct recently include the Pearson's hawthorn. It was last seen in 1994 in Mississippi, Louisiana, and Texas. In 1978, the Kingman's prickly pear was last seen in Arizona's Mojave Desert. In 1971, the Zion jimmyweed became extinct in Utah.

THE EFFECT OF INVASIVE SPECIES ON AGRICULTURE AND THE FOOD SUPPLY

Invasive species—often driven to migrate by climate change—can cause habitat destruction and an unbalanced ecosystem. But according to the National Wildlife Federation, they also cost billions of dollars each year to the United States economy. This is because invasive species can ruin crops, crowd out native species of plants and animals, and disrupt agriculture.

The loss of agriculture is more than just an economic issue. While it is true that farmers will lose millions of dollars because of invasive species, scientists predict that there will also be health and hunger issues to deal with as a result

This farmer sets a trap for brown marmorated stink bugs. The invasive species was first identified in Allentown, Pennsylvania, in 1998. It is now damaging large percentages of crops in thirty-three states.

of a threatened food supply. Despite the arrival of industri-alization, modernity, and digitization, humans still depend for survival upon the crops they grow. When crops become contaminated or compromised in some way, human health is at risk. Whether the problem is tainted foods that carry and spread disease or crop failure due to disease, drought, or flooding, the food supply will shrink and people will suffer.

The effects of climate change on agriculture and the food supply is potentially catastrophic. Unclean water, agricultural pests, and an increase in floods and severe weather events all have negative effects on crops. Right now we don't think about the possible extinction of the crops we rely upon for sustenance, but year after year of crop failures will put some plant species at risk. Disease can strike crops and spread

This Iowa farm lost over 100 acres (404,686 square meters) of farmland due to flooding in 2008.

through them quickly, wiping out a year's worth of work for farmers and potentially endangering the continued existence of certain crop varieties. In today's world, that would have consequences for more people than just the farmer. Many farm workers, truck drivers who transport the goods, stores that carry and decide on the price of the goods, and customers who buy the goods will be negatively impacted by large-scale disruptions in the size and health of the food supply.

CHAPTER four

Saving Species

Saving species from extinction is a serious business in the scientific field. Many zoos in the United States and around the world have opened centers that are dedicated to researching plant and animal ecosystems and how to preserve them. For example, the Wildlife Conservation Society works out of its headquarters at the Bronx Zoo, one of the leading zoos in the United States. The San Diego Zoo has a modern facility called the Conservation and Research for Endangered Species Center.

Other zoos work directly with teams of scientists in the habitats where species are in danger. They work with the public to educate them about forest birds in Hawaii and iguanas in the islands in the North Atlantic. They study the African wild dog in Zambia to see how it is adapting to its environment. The National Zoo in Washington, D.C., helps teach farmers in Africa how to prevent the spotted cheetah from ruining their crops without resorting to hunting. The Houston Zoo is working on saving the tapir in Venezuela, where it is endangered due to habitat loss and human activity.

Cutting-edge technology is also helping scientists to learn about species and the way they are adapting to the changing climate. This growing body of knowledge is helping to bring some endangered species back from the brink of extinction.

MONITORING SYSTEMS

Scientists know that the best way to monitor animals in their natural habitat is to do it without being detected. Since observing animals in the ocean is extremely difficult to begin with, remote video monitoring and satellite technology are great tools for observing ocean ecosystems.

Scientists at the Alaska SeaLife Center have used satellite devices and mounted them on or implanted them into marine mammals and sea birds. The devices transmit data about their location as well as their migration routes and where they find food. Over time this data will allow scientists to notice changes in these migration patterns or hunting habits. Combined with information about the water temperature,

Researchers sedate and blindfold a bear so that it can be tagged for tracking and monitoring by the Wildlife Conservation Society.

sea levels, and weather and climate changes, this data will help scientists get real data about the animals and how they are adapting to environmental changes. Some of these monitors provide a life history of up to a dozen years, including the cause of death of the animals.

Remote monitoring systems are ideal for scientists because they allow observation of the animals in a truly natural environment. The tools also provide records that can be collected over time and studied alongside data from other populations. As with other environmental studies about climate change, the remote satellite monitoring of species may take several years to produce useful data that can help scientists draw important conclusions about climate change. It is these conclusions that can help to make policy changes to safeguard the environment and provide financial support for further climate change studies.

CLONING ENDANGERED SPECIES

In 2001, scientists cloned the first endangered species. A year after a United States company cloned an Asian ox that died just forty-eight hours after birth, a team led by an Italian university successfully cloned a European mouflon. It is one of the world's smallest wild sheep. The animal lived originally on the islands of the Mediterranean but almost died out. The scientists collected cell samples from the sheep and other endangered animals. The hope was that animals could be preserved faster with cloning than they could by breeding the animals in captivity or having them be conceived and born naturally in the wild. With very endangered species,

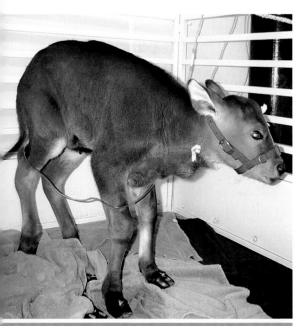

This Asian ox, called a gaur, was the first endangered species to be cloned by scientists. However, it died shortly after birth.

the rate of extinction is often faster than the rate at which the animals can reproduce.

Not all clonings are successful, however. The Pyrenean ibex represented an early failed attempt to clone an endangered animal species. Skin samples were taken from the last known Pyrenean ibex, and the clone was produced from this sample. However, the clone died of lung problems shortly after it was born, and the species became extinct.

Over the past couple of decades, there have been debates among scientists about whether it is ethical to clone animals. In the case of endangered species, some people say that cloning them gives humans a false sense that the animals are being preserved and that no further action is required to protect the environment and its various ecosystems. Some people may believe that if cloning is needed, then conservation methods have failed and the species will no longer be able to evolve naturally with genes from two parents or survive the changing conditions in its present habitat. These people may believe the cause is already lost and further conservation efforts would be wasted and doomed to failure.

EROSION AND HABITAT LOSS

One of the most potentially destructive consequences of global warming is the predicted rise in sea levels worldwide. Rising ocean levels will, in turn, lead to beach erosion and destruction of delicate and vital marshes and wetlands. Both beaches and wetlands are natural flood barriers, and wetlands also serve to filter and purify freshwater sources. Coastal areas, many of them large urban population centers, will be threatened with devastating flooding, and natural areas may suffer permanent damage to their plant, animal, fish, and bird habitats. The increasing number of coastal storms and severe weather events that are predicted to occur as global surface and ocean temperatures rise will only add to the coastal erosion problems. Wildlife in these areas will be put under more severe strain as coastal ecosystems are washed away.

Cloning skeptics also feel that we are not helping to improve the ecosystems of these animals by simply cloning them. We should instead be addressing the root problems of environmental degradation and climate change, and work to make the animals' ecosystem more balanced so that they can survive on their own. Others feel that preserving the species is the most important factor and that it should be done at any cost, even if that involves the relatively untested method of cloning.

Cloning has not yet taken off as the preferred method of preserving endangered species, but it is becoming a part of our daily reality. For example, scientists often discuss the idea of cloning plants that are resistant to disease. Cloning might also help to address world hunger problems, as plants that are the highest in nutrition or most drought- or pest-resistant may be chosen to be cloned. The plants that provide medicinal benefits might also be worth cloning. Some people claim that we are already consuming genetically altered plants when we enjoy seedless grapes, oranges, or watermelons that we can buy in the supermarket. Some plants, such as soybeans, are already altered to be resistant to certain pesticides.

USING TRADITIONAL CONSERVATION METHODS

Because scientific technology develops and changes over time, we will soon have more ways than ever to save endangered plant and animal species and to adapt to the inevitable extinctions of some of these organisms. Until then, however, scientists are working hard to use traditional conservation methods to save threatened and endangered species. These methods involve preserving land for plants and animals to live on without the disruption of human activity. Wildlife preserves are areas set aside for the flourishing of ecosystems and the preservation of food webs with a minimum of human interference or impact.

Another way that scientists and conservationists preserve species is by breeding endangered species in protected environments that exist apart from predators and human development. Following successful breeding, the animals are then reintroduced to the natural environment so they can live on their own without further human assistance. This was done for the bald eagle, which was close to extinction just a generation ago. The animal's habitat had been largely destroyed by humans. It had been hunted to the point of extinction because it was viewed as a threat to livestock. Even more damage was done to the bald eagle populations by a pesticide called DDT, which caused the birds to lay eggs with shells that were too thin to survive until the babies hatched.

The bald eagle was close to extinction, but conservation efforts eventually caused it to be removed from the endangered list.

Laws were put in place banning both hunting and the use of DDT. Nesting sites were protected so that the greatest number of eagles would be able to hatch. It took many years before populations of bald eagles returned in large enough numbers for the species to be removed from the endangered list. The return of the bald eagle is a success story that scientists wish would be the outcome of every conservation effort they attempt.

CHAPTER five

What You Can Do

When we think about the problem of climate change and global warming, it often seems too massive and complex a problem to successfully tackle or even begin solving. The crisis has been building since before we were born and will continue for centuries after our deaths. But there are plenty of things that people can do—especially young people—to help preserve and protect the planet. Even small steps that show that we are aware of how we should be living will help the environment and set a good example for others to be inspired by, follow, and build upon. Here are some things we can do to help keep ecosystems in balance for all living things, including humans.

REDUCE YOUR CARBON FOOTPRINT

When everyone becomes aware of the causes of climate change, we can work to prevent the problem from growing. Reducing your carbon footprint means being responsible for putting less greenhouse gases into the atmosphere because

Using less fuel is one way to reduce the amount of carbon dioxide released into the atmosphere. If it's safe, consider biking, walking, or taking public transportation when venturing out and about.

of daily activities. This can mean using alternative energy sources that are renewable, such as solar or wind energy. It can mean using cars that use alternatives to gasoline, such as electric hybrid vehicles. It can also mean driving only when needed, or taking public transportation, biking, or walking whenever possible. Even turning off the lights in your home or using electricity and electrical devices only when needed can help you dramatically reduce your carbon footprint (not to mention your electricity bill).

Many nations and corporations have made pledges to reduce their carbon emissions over time. If individuals and families did the same, then the population as a whole will be more environmentally aware and responsible. As the human population increases and developing nations begin to require and consume more energy and resources, it is even more necessary for people to use no more than what they need. Scientists agree that the reduction of greenhouse gases in the atmosphere is the only way to reduce the overall effects of climate change.

LEARN ABOUT PLANT AND ANIMAL POPULATIONS

Saving organisms from the brink of extinction is a difficult job for anyone, whether they are a trained scientist or con-servationist, or just an average citizen. But learning about endangered species is the first step in saving them. Do some research about the area you live in. What kind of endangered species are native to your area? What are the threats that are causing the populations to decline? What is being done to protect them? If you learn as much as possible about the species, you can better understand the problems facing it and get the word out to those around you.

There are many organizations that you can consult to learn more about the environment and get actively involved in its protection. The World Wildlife Fund, the Nature Conservancy, Greenpeace, and the Sierra Club are some of the leading environmental organizations that both educate the public and governments and work extremely hard on the ground to protect the environment, its fragile and threatened ecosystems, and all the life-forms that exist within them.

BECOME A VOLUNTEER

Some local branches of wildlife organizations have education programs that young people can get involved in. Cleaning local parks, wildlife areas,

Keeping plant and animal ecosystems pristine and reducing our negative impact upon them are good ways to help species not only survive but thrive as well.

lakes, and shorelines helps to make ecosystems safe for animals and makes soil healthy for plants to grow in. Sometimes, ocean communities are faced with oil spills that affect birds, animals, and plant life. Volunteers often help to clean these oil-coated animals and return the environment to a healthy condition in which animals can thrive again. But even picking up trash along a beach can help protect a seashore's animal and plant communities.

Starting school clubs or groups that raise awareness of endangered species and climate change is a good way to help out. Raising awareness can help change behaviors and get more people involved in helping to take care of the planet.

RESPECT THE ENVIRONMENT

You have learned about ways that plant and animal species, including humans, are affected by climate change. Respecting the environment is one of the best ways to make sure that you are helping to make a difference on behalf of endangered species. Be sure to follow any local laws about hunting or fishing. The laws are meant to protect the environment, preserve habitats, and maintain healthy animal populations. Avoid littering or damaging ecosystems with chemicals or pollution of any kind. Planting seeds of native plants helps to encourage plant growth while providing food for local animals. Respect the living things in your ecosystem, and remember that they have a role in the food chains and food webs that support your own environment, not to mention your own survival.

EXPEDITIONS FOR THE ENVIRONMENT

Teens who want to make a difference in the environment can join organizations specially designed for young people. The Earthwatch Institute is one such organization that teaches teens ages fifteen to eighteen about environmental concerns and goes on expeditions around the world to learn about the effects of climate change. Participants can travel to and discover habitats such as rain forests, ocean ecosystems, and the Arctic. These teams of teens take part in actual field research as part of their education, and this research adds to the body of knowledge that is allowing scientists to come to grips with climate change and its effects upon animal and plant species.

SUPPORT ENVIRONMENTAL LEGISLATION

It is important to be informed about how local, state, and federal politicians feel about environmental issues. When they support legislation that will protect the environment or help slow the effects of climate change, people have to be aware of it, express their approval of pro-environment laws, and support the election and reelection campaigns of pro-environment candidates. Find out where leaders stand on the issues and consider supporting the ones that vow to be most helpful for the environment.

BE AWARE OF INVASIVE SPECIES

Invasive species can cause ecosystems to be greatly damaged. They have even caused the extinction of some plant

When camping, use only firewood from inside the park or immediate locality to cut down on the spread of invasive insect species that could be living in wood brought in from elsewhere.

and animal species. Be aware of species that do not belong in an ecosystem. Plant only native plants in your yard or garden and remove invasive plants from your property. Report to your local department of natural resources any new species of plants or animals that you have never seen before in your area.

When camping, use firewood only from the area you are camping in. Many insects and their eggs travel in firewood and are introduced to a new area when the wood is relocated. Insects can cause damage to an ecosystem by competing for resources with native species. When they do not have natural predators in the area, they tend to take over and cause an imbalance in the food webs and ecosystem.

GLOSsary

biodiversity The variety of life in any ecosystem or area.

carbon dioxide A colorless, odorless gas that is a natural component of the atmosphere and is also released when fossil fuels are burned.

carbon footprint The amount of carbon dioxide released due to the activities of a nation, state, community, group, or person.

climate change A long-term change in Earth's climate, sometimes as a result of human activity and driven by global warming.

climate model A computer model that predicts future climate conditions over a period of time.

clone A genetic replica of an organism made from the cells of that organism.

conservationist A person who works for the protection of the environment and wildlife.

coral bleaching The loss of color of coral reefs due to the loss of algae.

ecosystem A biological environment consisting of all the organisms living in a particular area, as well as all the nonliving, physical components of the environment with which the organisms interact, such as air, soil, water, and sunlight.

emission The production or release of something, such as a gas.

endangered In terms of plant and animal species, being in imminent danger of becoming extinct.

erosion The wearing away of Earth's soil or sand through natural processes, such as wind and water flow.

extinct The state of having no more living members of a particular species on Earth.

fossil fuel Natural fuel formed over centuries from the remains of living organisms; includes coal, oil, and natural gas.

global warming A long-term increase in average global surface and ocean temperatures, largely as a result of human activity; a key driver of climate change.

greenhouse effect The trapping of the sun's warmth in Earth's lower atmosphere.

greenhouse gases Gases that contribute to the greenhouse effect, particularly carbon dioxide and methane.

invasive species A non-native organism introduced to a new area where it often has no natural predators and can begin to compete with and crowd out native species.

legislation Bills that are proposed, debated, voted upon, and passed into law.

threatened In terms of plant and animal species, being vulnerable to endangerment in the near future; in danger of becoming extinct.

Cities for Climate Protection
15 Shattuck Square, Suite 215
Berkeley, CA 94704
(510) 540-8843
Web site: http://www.iclei.org/co2
Cities for Climate Protection promotes reductions in green-
  house gas emissions of towns and cities across the
  United States.

Clean Air Council
135 South 19th Street, Suite 300
Philadelphia, PA 19103
(215) 567-4004
Web site: http://cleanair.org
The Clean Air Council is a nonprofit environmental education and advocacy organization. Based in Pennsylvania,
  the organization provides information about carbon
  emissions and climate change.

Climate Action Network Canada
412-1 Nicholas Street
Ottawa, ON K1N 7B7
Canada
(613) 241-4413
Web site: http://www.climateactionnetwork.ca
Climate Action Network Canada is an organization of members committed to preventing human interference in the
  global climate. It also promotes renewable energy as a
  method of reducing greenhouse gas emissions.

Environmental Protection Agency (EPA)
1200 Pennsylvania Avenue NW
Washington, DC 20460

(202) 272-0167
Web site: http://www.epa.gov
The EPA provides information to the public about environmental issues such as climate change and how to
reduce carbon emissions.

Environment Canada
10 Wellington, 23rd Floor
Gatineau, QC K1A 0H3
Canada
(819) 997-2800
Web site: http://www.ec.gc.ca
Environment Canada is a department of the Canadian gov-
ernment devoted to protecting the environment and the
people of Canada. It has conducted substantial research
and presented a number of publications on greenhouse
gas emissions, global warming, climate change, and
plant and animal extinctions.

Greenpeace
702 H Street NW, #300
Washington, DC 20001
(202) 462-1177
Web site: http://www.greenpeace.org
Greenpeace is a conservation organization founded in
Vancouver, Canada, that is dedicated to exposing
threats to our natural resources and planet.

Nature Conservancy
4245 North Fairfax Drive, #100
Arlington, VA 22203-1606
(703) 841-5300

Web site: http://www.nature.org
The Nature Conservancy is an environmental organization
    dedicated to preserving the land and water that support
    biodiversity on Earth.

Sierra Club
82 Second Street, 2nd Floor
San Francisco, CA 94105
(415) 977-5500
Web site: http://www.sierraclub.org
The Sierra Club is the largest environmental organization in
    the United States.

World Wildlife Fund
1250 24th Street NW
P.O. Box 97180
Washington, DC 10090
(202) 293-4800
Web site: http://www.worldwildlife.org
The World Wildlife Fund is a conservation organization that
    seeks to protect nature and reduce threats to
    biodiversity.

WEB SITES

Due to the changing nature of Internet links, Rosen Publishing
has developed an online list of Web sites related to the sub-
ject of this book. This site is updated regularly. Please use
this link to access the list:

http://www.rosenlinks.com/sttr/extin

FOR FURTHER READing

Bailey, Gerry. *Changing Climate*. New York, NY: Gareth Stevens Publishing, 2011.

Baillie, Jonathan, and Marilyn Baillie. *Animals at the Edge: Saving the World's Rarest Creatures*. Toronto, ON, Canada: Maple Tree Press, 2008.

Benoit, Peter. *Climate Change: A True Book*. New York, NY: Scholastic Library Publishing, 2011.

Bradman, Tony. *Under the Weather: Stories About Climate Change*. London, England: Frances Lincoln Children's Books, 2010.

Eyewitness Books. *Climate Change: Eyewitness Guide*. New York, NY: Dorling Kindersley, 2011.

Hanel, Rachael. *Polar Bears*. Mankato, MN: The Creative Company, 2009.

Jenkins, Martin. *Can We Save the Tiger?* Somerville, MA: Candlewick Press, 2011.

Johnson, Rebecca L. *Investigating Climate Change: Scientists' Search for Answers in a Warming World*. Minneapolis, MN: Twenty-First Century Books, 2008.

Maczulak, Anne E. *Biodiversity: Conserving Endangered Species*. New York, NY: Facts On File, 2009.

Marcovitz, Hal. *How Serious a Threat Is Climate Change?* San Diego, CA: Reference Point Press, Inc., 2011.

McCutcheon, Chuck. *What Are Global Warming and Climate Change? Answers for Young Readers*. Albuquerque, NM: University of New Mexico Press, 2010.

Nagle, Jeanne M. *Endangered Wildlife: Habitats in Peril*. New York, NY: Rosen Publishing, 2009.

Ollhoff, Jim. *Climate Change: Living in a Warmer World*. Mankato, MN: ABDO Publishing, 2010.

Ollhoff, Jim. *Climate Change: Myths and Controversies*. Mankato, MN: ABDO Publishing, 2010.

Oxlade, Chris. *Climate Change* (Science in the News). London, England: Franklin Watts, 2008.

Rafferty, John P. *Climate and Climate Change*. New York, NY: Britannica Educational Publishing, 2011.

Sheehan, Sean. *Endangered Species*. New York, NY: Gareth Stevens Publishing, 2009.

Sivertsen, Linda. *Generation Green: The Ultimate Teen Guide to Living an Eco-Friendly Life*. New York, NY: Simon Pulse, 2008.

Solway, Andrew. *Climate Change* (World at Risk). New York, NY: W. B. Saunders, 2011.

Spilsbury, Richard. *Ask an Expert: Climate Change*. New York, NY: Crabtree Publishing Company, 2010.

Taylor, B. *Planet Animal: Saving Earth's Disappearing Animals*. Hauppauge, NY: Barron's Educational Series, 2009.

Taylor, Nancy H. *Go Green: How to Build an Earth-Friendly Community*. Layton, UT: Gibbs Smith, 2008.

Walker, Sally M. *We Are the Weather Makers: The History of Climate Change*. Somerville, MA: Candlewick, 2009.

BIBLIOGraphy

Alaska SeaLife Center. "Cutting Edge Technology." Retrieved October 2011 (http://www.alaskasealife.org/New/research/index.php?page=cutting_edge.php).

Cevallos, Marissa. "Corals Moving North." *U.S. News & World Report*, January 24, 2011. Retrieved July 2011 (http://www.usnews.com/science/articles/2011/01/24/corals-moving-north?s_cid=related-links:TOP).

Cimons, Marlene. "A Center to Study Oceans and Climate." *U.S. News & World Report*, September 27, 2010. Retrieved July 2011 (http://www.usnews.com/science/articles/2010/09/27/a-center-to-study-oceans--climate?s_cid=related-links:TOP).

Conniff, Richard. "How Species Save Our Lives." *New York Times*, February 27, 2011. Retrieved September 2011 (http://opinionator.blogs.nytimes.com/2011/02/27/how-species-save-our-lives/).

Currentresults.com "Recently Extinct Animals and Plants in North America." Retrieved September 2011 (http://www.currentresults.com/Endangered-Animals/North-America/recently-extinct-animals.php).

Earthwatch.org "Expeditions for Teenagers Aged 15–18." EarthWatch Institute. Retrieved October 2011 (http://www.earthwatch.org/expedition/teenteam).

Eurekalert.org. "Coping with Climate Change." May 11, 2011. Retrieved September 2011 (http://www.eurekalert.org/pub_releases/2011-05/nesc-cwc051111.php).

Gerstein, Julie. "Extinct in Our Lifetime—11 Species We've Lost Forever." The Daily Green. Retrieved September 2011 (http://www.thedailygreen.com/environmental-news/latest/recently-extinct-animals-list-470209).

Hillman, Mayer, Tina Fawcett, and Sudhir Chella Rajan. *The Suicidal Planet*. New York, NY: St. Martin's Press, 2007.

Hoggan, James. *Climate Cover-Up: The Crusade to Deny Global Warming*. Vancouver, BC, Canada: Greystone Books, 2009.

Hooper, Meredith. *The Ferocious Summer: Adélie Penguins and the Warming of Antarctica*. Vancouver, BC, Canada: Greystone Books, 2008.

International Business Times. "Climate Change May Impact Extinction Risk of Animal Populations: Study." April 6, 2011. Retrieved September 2011 (http://www.ibtimes .com/articles/131113/20110406/climate-change-environment-global-warming-animal-populations-spectral-colour.htm).

Milius, Susan. "Acidification May Halve Coral Class of 2050." *U.S. News & World Report*, November 10, 2010. Retrieved 2011 (http://www.usnews.com/science/ articles/2010/11/10/acidification-may-halve-coral-class-of-2050?s_cid=related-links:TOP).

Miller, Amy. "What Zoos Do." Scholastic.com. Retrieved October 2011 (http://teacher.scholastic.com/ scholasticnews/indepth/endangered_species/zoos/ index.asp?article=whatzoosdo).

National Wildlife Federation. "Global Warming and Polar Bears." Retrieved September 2011 (http://www.nwf.org/ Global-Warming/Effects-on-Wildlife-and-Habitat/ Polar-Bears.aspx).

National Wildlife Federation. "Invasive Species." Retrieved September 2011 (http://www.nwf.org/Wildlife/ Wildlife-Conservation/Threats-to-Wildlife/Invasive-Species.aspx).

ScienceDaily.com "Climate Change Played Major Role in Mass Extinction of Mammals 50,000 Years Ago, Study Finds." May 18, 2010. Retrieved

September 2011 (http://www.sciencedaily.com/
releases/2010/05/100518064614.htm).

ScienceDaily.com "Thousands of Undiscovered Plant
Species Face Extinction Worldwide." July 7, 2010.
Retrieved September 2011 (http://www.sciencedaily
.com/releases/2010/07/100707065218.htm).

Strain, Daniel. "Extinctions Breed Carbon Chaos." *U.S.
News & World Report*, February 7, 2011. Retrieved July
2011 (http://www.usnews.com/science/articles/2011/
02/07/extinctions-breed-carbon-chaos?s_cid=
related-links:TOP).

Telegraph. "Top 10 Most Endangered Species in the
World." January 4, 2010. Retrieved September 201.
(http://www.telegraph.co.uk/earth/wildlife/6927330/
Top-10-most-endangered-species-in-the-world.html).

Trivedi, Bijal P. "Scientists Clone First Endangered Species:
a Wild Sheep." *National Geographic*, October 20, 2001.
Retrieved October 2011 (http://news.nationalgeographic
.com/news/2001/10/1025_TVsheepclone.html).

University of Texas at Austin. "Global Warming Increases
Species Extinction Worldwide, University of Texas
at Austin Researcher Finds." November 14, 2006.
Retrieved September 2011 (http://www.utexas.edu/
news/2006/11/14/biology/).

U.S. Fish & Wildlife Service. "What Is the Difference
Between Endangered and Threatened?" Retrieved
September 2011 (http://www.fws.gov/midwest/wolf/
esastatus/e-vs-t.htm).

U.S. News & World Report. "Climate Change Is Making Our
Environment 'Bluer.'" April 7, 2011. Retrieved July 2011
(http://www.usnews.com/science/articles/2011/04/07/
climate-change-is-making-our-environment-bluer).

U.S. News & World Report. "Greenhouse Ocean Study Offers Warning for Future." May 19, 2011. Retrieved September 2011 (http://www.usnews.com/science/articles/2011/05/19/greenhouse-ocean-study-offers-warning-for-future?s_cid=related-links:TOP).

Watson, Traci. "86 Percent of Earth's Species Still Unknown?" *National Geographic*, August 24, 2011. Retrieved October 2011 (http://news.nationalgeographic.com/news/2011/08/110824-earths-species-8-7-million-biology-planet-animals-science/).

INDex

A

Arctic, and melting sea ice, 4, 5, 13

B

bald eagle, conservation efforts and, 43
breeding programs, 42–43

C

carbon dioxide, 10–12, 13, 20, 27
carbon footprint, reducing your, 45–46
climate change, 6
  causes of, 7–18
  predicted results of, 22–24, 25, 35
  predictions for future, 19–28
cloning endangered species, 39–42
computer models for climate change,
    19–20, 25
  data collected for, 20–21
conservation efforts, 18, 36–43
  using traditional, 42–43
  what you can do, 44–50
coral bleaching, 28

E

endangered animals and plants
  cloning, 39–42
  definition of, 16
  examples of, 16–18
  learning about, 46–47
Endangered Species Act, 18
environment, respecting the, 48
erosion, habitat loss and, 41
expeditions, environmental, 49

extinctions
  consequences of, 29–35
  current rate of, 4, 12, 31
  definition of, 15
  human role in, 14
  notable, 32–33
  periods of mass, 6
  predictions for, 12, 22–24
  saving species from, 36–43

F

food chain/web, effects of
    being altered, 6, 13, 22, 23,
    25–27, 28
fossil fuels, burning of, 8–10, 12

G

global warming, 11–12
  effect of, 12–14, 41
  and increasing rates of extinction,
    4, 12, 25
  and prospects for plants and animals,
    21–22, 27–28
greenhouse effect, 11–12
greenhouse gases, 10, 11, 12, 20,
    21, 45, 46
  legislation to decrease, 16

H

habitat destruction, 14, 18, 24,
    32, 33, 37, 41
human activity
  as cause of climate change, 7,
    8–10, 12
  and extinctions, 14, 18, 24, 32–33

ABOUT THE AUTHOR

Kathy Furgang has written many books for young readers, including numerous textbooks about science and the environment. The topics she has written about include weather, climate change, economics, and history. She lives in upstate New York with her husband and two sons.

PHOTO CREDITS

Cover, pp. 1, 3, 4-5 (background) Tom Brakefield/Stockbyte/ Getty Images; p. 5 © istockphoto.com/Josef Friedhuber; p. 9 Welgos/Getty Images; p. 11 Monica Schroeder/Science Source/Photo Researchers, Inc.; p. 15 George Rose/Getty Images; p. 17 Eric Lafforgue/Gamma-Rapho/Getty Images; pp. 20–21 DPA/Newscom; p. 24 © istockphoto.com/brytta; pp. 26, 28, 34, 50 © AP Images; p. 30 Inga Spence/Visuals Unlimited; p. 35 David Greedy/Getty Images; p. 38 Steve Winter/Contributor/National Geographic; p. 40 Advanced Cell Technology/Newsmakers/Getty Images; p. 43 Daniel Acker/Bloomberg/Getty Images; p. 45 Hemera/Thinkstock; p. 47 Michael Buckner/Getty Images; interior background (globe) © istockphoto.com/m-a-r-i; back cover (world), interior background chudo-yudo/Shutterstock.com.

Designer: Nicole Russo; Photo Researcher: Amy Feinberg